INTRODUCTION TO FAMILY AND CONSUMER SCIENCES (FCS)

University of Montevallo
Montevallo, AL

innovativeink
PUBLISHING
A Division of Kendall Hunt

Cover images Shutterstock.com

innovativeink PUBLISHING
A Division of Kendall Hunt

www.innovativeinkpublishing.com
Send all inquiries to:
4050 Westmark Drive
Dubuque, IA 52004-1840

Print ISBN: 979-8-3851-9410-0
Ebook ISBN: 979-8-3851-9412-4

Published in the United States of America

Contents

Course Description

The course will focus on the Family and Consumer Sciences body of knowledge and assist students in understanding their roles as Family and Consumer Sciences professionals.

Students in this course will:

1. Analyze the impact of personal and interpersonal relationships of families and individuals on home, work and other environments.
2. Understand the dynamics of capacity building of individuals within families, communities, work environments, and other contexts.
3. Understand their roles as Family and Consumer Sciences professionals and support the well-being of individuals, families, consumers, and communities.
4. Develop critical thinking skills to utilize diverse approaches for determining alternative solutions for issues.
5. Understand the principles and importance of building collaborations and partnerships on campus and in communities and in the workplace and apply those principles to various situations.
6. Demonstrate knowledge of potential careers, professional associations, and credentials offered in the various areas of Family and Consumer Sciences; develop basic skills for successful entrance into the Family and Consumer Sciences field as a professional.

Jimmy Rooney/Shutterstock.com

1

Welcome

GLOSSARY

ACADEMIC ADVISING. Students meet with their academic advisor at least once a semester to discuss progress towards completion of their degree. Advising is a time for students and advisors to explore potential scholarship opportunities, continuing education, internship possibilities, and general support for the overall well-being of the student.

ACCREDITATION. Accreditation applies to institutions or programs and is distinguished from certification or licensure, which applies to individuals. According to the Council for Higher Education (CHEA), accreditation is "a collaborative and collegial review process, is based on self and peer-assessment for public accountability and continual improvement of academic quality."

BODY OF KNOWLEDGE. The subject matter of Family and Consumer Sciences forms the basis of accreditation by the Council for Accreditation. From the perspective of a holistic and synergistic focus, Family and Consumer Sciences professionals relate to the interrelationships among individuals, families, and communities through (a) their study of human systems theory and life course development, (b) capacity building, (c) issues that foster wellness, (d) global interdependence, (e) appropriate use of technology, (f) resource development and sustainability, and (g) public policy development and analysis. An understanding of the complexity of the reciprocal relationships (interrelationships and interdependence) among individuals, families, communities, and the many environments in which they function throughout their life spans is the unifying focus of the Family and Consumer Sciences knowledge base. An understanding of individual, family, and community needs, ecosystems, and life course development underpins the dynamic, holistic, and integrative nature of Family and Consumer Sciences. (AAFCS.org)

CANVAS. The University utilizes Canvas as a learning management system. Faculty utilization of Canvas varies by course. Faculty can utilize Canvas to post announcements, send emails, post course syllabus, post course documents and assignments, etc.

CAPACITY BUILDING. The practice of individuals, families, and communities acquiring knowledge and skills, building on assets and strengths, respecting diversity, and responding effectively to change. Some of the competencies required for 21st Century living include the ability of human beings to understand and solve complex problems, to preserve the knowledge of indigenous cultures in the context of rapid globalization, to use technology appropriately, and to strive for balance between change and continuity. Institutional capacity depends upon the competencies of individuals and families and their commitment to build sustainable communities. (AAFCS.org)

COURSE OBJECTIVES. What students can expect to learn and take away from the course.

CREDIT HOUR. Credit hours are assigned to courses based on the expected level of content and learning involved. Generally, a one credit hour course is equivalent to one hour of class meeting per week with an additional two hours of out of class work. The University of Montevallo requires a minimum of 120 credit hours of required course work per major for graduation. Individual majors may have slightly higher graduation requirements.

DEGREEWORKS. DegreeWorks is an online system used by students, advisors, faculty and administration to track progress towards degree completion. It provides a clear view of completed and required coursework.

FAMILY SYSTEMS. Family systems are comprised of the interrelated elements of family members, each with his or her own independence, yet being interdependent upon the other members of the family. (AAFCS.org)

GLOBAL INTERDEPENDENCE. The impact of people individually and collectively on the world and, in turn, the effect of their actions upon others. (AAFCS.org)

HUMAN SYSTEMS THEORY. A theory that seeks to explain human development and behavior in the context of, and in interaction with, various environments. (AAFCS.org)

INTEGRATIVE FOCUS. The unique concepts that address the well-being of individuals, families, consumers, and communities and verify the disciplines within the Family and Consumer Sciences profession. (AAFCS.org)

LIBERAL ARTS EDUCATION. Liberal arts education encourages students to think critically, communicate effectively, adapt to change, and become well-rounded citizens. As students engage in a liberal arts curriculum, they explore the world through different lenses and become knowledgeable of different methods of interpreting, experiencing, understanding, and analyzing the world around them. Students are encouraged to think broadly and become problem solvers by discerning which problems to solve and why. A liberal arts education helps students succeed in the classroom and prepares them to tackle some of the world's most complicated issues.

MISSION STATEMENT. A mission statement defines the core purpose, guiding principles, and overall goals of a university, college, and or department. The mission statement is a roadmap for communicating values, operations, contributions to society, and research.

ORGANIZATION OF THE UNIVERSITY

- Colleges (College of Education and Human Development; College of Fine Arts; College of Health Professions; College of Arts and Sciences; College of Business): Within the broader university, colleges are specialized academic units focused on particular fields of study.

- Departments (Within each college, degree programs may be grouped into departments)
- Programs (Programs are specific majors under each department).
- Majors/Concentrations (The main area of study a student chooses to specialize in).
- **President:** Responsibility for the total operation of the University rests with the President of the University as the chief executive officer of the University.
- **The Provost and Senior Vice President for Academic Affairs (Provost/Sr. VPAA)** reports to the President and is the principal advisor on academic matters and, in the President's absence, serves as Acting President.
- **Academic Deans** report to the Provost and are the administrator over each college providing leadership in curriculum offerings and instruction.
- **Academic Department Chairs** report to the Dean of the College and are responsible for management of the department to include recruitment, supervision and evaluation of faculty and staff; periodic curricular review and revision; budget and facilities management; and program assessment.
- **Academic Program Coordinators** report to the Department Chair and works with program faculty, the Chair and the Dean to ensure that program curricula are current; are aligned with institutional mission and accepted standards of practice and/or scholarship; are responsive to student needs; and are approved through the established curriculum governance/approval process.
- Professor/associate professor/assistant professor / instructor is the title held by university faculty.
  - An instructor is an educator who teaches and guides students within a degree program; an instructor is not a tenure track position.
  - An assistant professor is an entry-level tenure track position; responsibilities include teaching, research, and service.
  - An associate professor indicates more experience and is usually tenured, indicating long-term commitment to the university; responsibilities include teaching, research, and service.

- A professor is the highest faculty rank and indicates the most senior and experienced faculty members: responsibilities include teaching, research, service, and leadership.
- An adjunct instructor / professor is a part-time educator that has specialized education in a specific field.
- Advisors are university faculty or staff that guide students toward degree completion in their individual plan of study, tracking progress each semester and offering support.

PUBLIC POLICY DEVELOPMENT AND ANALYSIS. Public policy analysis is the systematic evaluation of government laws, regulations, and programs to understand how they impact individuals, families, and communities. This analysis is used to advocate for or develop policies that improve quality of life, promote equity, and support sustainable and healthy living environments for diverse populations.

SUSTAINABLITY The term refers to meeting the needs of the present without comprising the ability of future generations to meet their own needs, involves consideration of the entire life cycle of people, buildings, the natural environment, and other components regarding economics, environmental impact, and performance. (aafcs.org)

SYLLABUS. A college course syllabus is your guide for the expectations of being a student in the course. The syllabus will include faculty details, faculty office hours, course description, course objectives, course policies (attendance, grading, late work, make-up exams, etc.), schedule of assignments and exams, and required materials and technology. The syllabus will have University policies listed or a link to those policies. University policies are listed in the University of Montevallo Bulletin. Students should familiarize themselves with all University policies.

WELLNESS. Wellness addresses health and well-being. It includes food security; adequate nutrition; reduced risk of chronic and communicable disease; access to forms of exercise; respectful, caring, and compassionate learning environments; healthcare availability, access, costs, and quality; psychological health; protection from abuse, exploitation, and violence; access to safe water and air and adequate sanitation facilities; control of healthcare costs so quality care is available; and spirituality. (aafcs.org)

taniascamera/Shutterstock.com

WELCOME TO FCS! UNIVERSITY OF MONTEVALLO

Welcome

Welcome to the Family and Consumer Sciences (FCS) program at the University of Montevallo. The program is under the College of Education and Human Development and the Department of Consumer and Human Sciences.

Much like the theme of the board game LIFE, FCS examines a person's journey throughout their life, from "womb to the tomb". It focuses on the science and art of living and working well in our complex world. Areas of focus include the culinary arts, food science and nutrition, event planning, wellness, housing, infant/child/and human development, family relations, marriage, human sexuality (yes, the birds and the bees), personal and family finance, textiles, apparel, and retail. FCS is all about LIFE. We look forward to embarking on this journey with you.

PURPOSE OF THE FCS FIELD

mathom/Shutterstock.com

As a people-centered science, the discipline of Family and Consumer Sciences is focused on the study of individuals, families, and communities through education, research, and technology. The Family and Consumer Sciences discipline provides individuals with the necessary knowledge and skills to live and work in a diverse and globally connected society.

Businesses of all kinds need individuals knowledgeable about consumer behavior, and Family and Consumer Sciences professionals have a history of working with businesses to support their work.

Topics that are covered in the Family and Consumer Sciences area of study include:

- Child development
- Consumerism
- Consumer safety
- Culinary arts
- Education and training
- Elder services
- Family relations
- Fashion design and merchandising
- Financial literacy
- Food science
- Hospitality and tourism
- Individual and family development
- Nutrition
- Retail merchandising
- Textiles
- Visual merchandising

FAMILY AND CONSUMER SCIENCES STUDENT LEARNING OUTCOMES

The unique focus of the Family and Consumer Sciences program is on families, work, and their interrelationships. The Family and Consumer Sciences program has determined that upon successful completion of the Family and Consumer Sciences major, students will be able to:

- Evaluate evidence -based family and consumer research and apply scientific reasoning for family and consumer interests.
- Use professional skills for effective communication, team building, and decision making consistent with standards of ethical practice.
- Identify relevant issues in the legislative process of public policy as it relates to Family and Consumer Sciences.
- Become responsible citizens and leaders in family, community, and work settings.
- Apply critical and creative thinking skills in addressing individual, family, and work-related problems and issues in diverse environments.
- Demonstrate the ability to use knowledge, skills, competencies, and attitudes in professional work experience.
- Create a professional portfolio.

MISSION STATEMENT

MISSION STATEMENTS

a. **Mission Statements and Philosophy**

University of Montevallo (reaffirmed 2021)

The overriding mission of the Universi-

Jimmy Rooney/Shutterstock.com

ty of Montevallo, unique in Alabama higher education, is to provide to students from throughout the state an affordable, geographically accessible, "small college" public higher educational experience of high quality with a strong emphasis on undergraduate liberal studies and with professional programs supported by a broad base of arts and sciences, designed for their intellectual and personal growth in pursuit of meaningful employment and responsible, informed citizenship.

College of Education and Human Development
The mission of the College of Education and Human Development is to prepare highly skilled, knowledgeable, and continually developing undergraduate and graduate candidates who make informed decisions, use current evidence and research-based practices, and promote healthy and equitable communities through careers in counseling, Family and Consumer Sciences, leadership, teaching, and technology.

American Association of Family and Consumer Sciences (updated spring 2023)
AAFCS elevates the profession by inspiring research, leadership, and service to empower individuals, families, and communities.

UM Family and Consumer Sciences Mission Statement (updated Jan. 2024)
The Family and Consumer Sciences (FCS) major prepares students to make positive contributions to the well-being of individuals, families, and consumers, through professional careers in education, business and industry, community service, government, and science and technology.

We strive to prepare professionals to be responsible and informed citizens, and for an increasingly diverse society by providing a range of experiences and opportunities that support their knowledge and skills. Capstone courses provide experiential learning and networking opportunities.

The mission of the program clearly reflects the general philosophy of the Family and Consumer Sciences profession. There is an emphasis on the practical application of reflective knowledge designed to improve well-being, relationships, and resources among individuals and families in an effort to achieve optimal functioning and quality of life.

Philosophical foundations and knowledge specific to the Family and Consumer Sciences profession address ongoing changes and major trends in society, families, and education. Current evidence suggests that society is faced with a number of perennial issues that negatively impact individuals, families, and communities. The mission of the unit is to utilize current research and evidence-based practices that will promote well-being in a contemporary society and improve outcomes and trajectories for individuals and families. A focus on content including nutrition and wellness; child and human development; interpersonal relationships; communication, diverse families; as well as the contemporary and historical ways of meeting the economic, physiological, psychological, and sociological needs of consumers relative to apparel and textile products is significant.

The integrating focus of the Family and Consumer Sciences program is the enhancement of interpersonal relationships of people with their environment. The Family and Consumer Sciences mission integrates research and service into the educational experience to address contemporary practice in the profession and in specialized career fields.

The Family and Consumer Sciences program and mission here at the University of Montevallo is distinct in that the program provides accredited, integrative, quality, and professional programs in specialized areas related to the well-being of individuals and families within a diverse and global society. The integrating focus of the unit is to develop graduates who effectively apply knowledge of interacting relationships between people and their environments, and to empower, strengthen, and enable individuals, families, and communities to thrive.

The Family and Consumer Sciences program mission promotes the vision and goals of the University of Montevallo and the College of Education and Human Development, which also emphasizes overall optimal functioning of individuals and communities, and seeks to prepare graduates who are skilled, knowledgeable and committed to professional development in their specialized field. The University is a liberal arts institution committed to its mission of providing a high-quality educational experience in professional programs supported by arts and sciences and liberal studies. The Family and Consumer Sciences program provides professional programs that have approximately 40% of the course work grounded in courses defined traditionally as liberal arts courses. Additionally, the program promotes the liberal arts focus of the University by providing three courses that have been approved for general education credit. The Family and Consumer Sciences program is accredited by the American Association of Family and Consumer Sciences (AAFCS), which aims to advance academic quality, demonstrate accountability, and encourage, where appropriate, planning and implementation for needed improvement. The University has a strong emphasis on undergraduate liberal art studies. Approximately 86% of the enrollment in the University is focused on undergraduate programs.

MAJORS AND MINORS

University of Montevallo and the Family and Consumer Sciences Program.

Family and Consumer Sciences is one of two programs within the Department of Consumer and Human Sciences in the College of Education and Human Development. The Family and Consumer Sciences degree is one undergraduate major, Family and Consumer Sciences. Students have the opportunity to select the Family and Consumer Sciences major without a concentration or select one of four concentrations included in the curriculum: Family and Consumer Sciences education, hospitality and tourism, human development and family studies, and retail merchandising. Students in the Family and Consumer Sciences major may select either a Bachelor of Science or a Bachelor of Arts degree program. Minors are available in human development and family studies, nutrition, and retail merchandising.

B.S./B.A.

Students in the Family and Consumer Sciences program, regardless of concentration, have the choice to choose between a Bachelor of Science and a Bachelor of Arts degree. A Bachelor of Science degree requires additional credit hours in mathematics and science. A Bachelor of Arts degree requires additional credit hours in foreign language courses.

Family and Consumer Sciences Education

Family and Consumer Sciences Education prepares students for high demand teaching positions in Alabama and across the United States. Individuals with a Family and Consumer Sciences education degree are equipped with the knowledge and skills to work as secondary education teachers (middle and high schools), career and technical education teachers, cooperate extension agents, corporate trainers, and curriculum developers to name a few career paths.

When individuals complete the Family and Consumer Sciences education degree, they will have the opportunity to take the ServSafe examination to become ServSafe certified. The ServSafe certification is recognized by the restaurant and food industry and recognized by many federal, state, and local jurisdictions. (ServSafe.com)

Air Images/Shutterstock.com

Anthony Mooney/Shutterstock.com

"The ServSafe Food Safety Training program leads the way in providing current and comprehensive educational materials to the restaurant industry. More than 4 million foodservice professionals have been certified through the ServSafe Food Protection Manager Certification Exam, which is accredited by the American National Standards Institute (ANAB) Conference for Food Protection (CFP). (ServeSafe.com)

Hospitality and Tourism

Hospitality and Tourism trains people to create and maintain spaces that make consumers feel welcome. A foundational aspect of the hospitality and tourism disciple is understanding how to take care of the customer. A hospitality and tourism degree provides individual the knowledge and skills to enter careers such as event management, food service operations, hotel and resort leadership, conferences and conventions, parks and recreations, sporting events and many others.

Krakenimages.com/Shutterstock.com

Human Development and Family Studies

Human Development and Family Studies (HDFS) prepares students to address real-world problems facing individuals and families in a changing and diverse global society. An HDFS degree prepares students for work in various fields including, but not limited to, human resources, childcare, youth development, policy analysis, family advocacy, and family life education.

Lazy Baboon/Shutterstock.com

Retail Merchandising

Retail Merchandising prepares students to work in fashion merchandising, retailing, and other areas of the apparel and textile industry by learning knowledge and skills to prepare them for the consumer-oriented business practices in buying, public relations, retail marketing and management,

DC Studio/Shutterstock.com

fashion forecasting, textiles, sales, visual merchandising and other essential position that are critical to this creative industry.

Certifications

Sergio Vas/Shutterstock.com

Certified Family Life Educator

Family Life Education is a field which aims to equip family members with the knowledge and skills to enhance well-being, strengthen relationships, and foster positive individual and family development across the lifespan (NCFR.org).

Rido/Shutterstock.com

How to Become a Certified Family Life Educator

Recent graduates of the University of Montevallo HDFS program can apply for *provisional* CFLE certification.

- Each CFLE-approved program has a checklist of coursework that has been approved by NCFR to meet the CFLE requirements. Students must complete all courses on the school checklist with grades of C- or better (in addition to an internship/practicum of at least 120 hours) to qualify for the CFLE-approved program application process:

NATIONAL COUNCIL ON FAMILY RELATIONS

ncfr catalyzing research, theory and practice

CFLE-Approved Program Checklist

Provisional Certified Family Life Educator (CFLE)

University of Montevallo, Undergraduate

B.A. & B.S., Family & Consumer Sciences

NCFR-approved program since 2014

Submit this checklist with all other materials required for the CFLE-approved program (abbreviated) application process. Find information and forms at: https://www.ncfr.org/cfle-approved-program

Instructions: Confirm that the courses and internship/practicum experience listed below appear on your transcript by placing a checkmark in the box to the left of the content areas. For any course substitutions (maximum two), include course syllabi from the year you completed the course. Use the blank row at the bottom to indicate content area and course title.

✓	CFLE Content Area	CFLE-Approved Program (Abbreviated) Coursework
	1. Families & Individuals in Societal Contexts	**Prior to Summer 2025:** FCS 402 - Dynamics of Family Relationships **AND** FCS 291 - Individual & Family Development **Summer 2025 onward:** FCS 402 - Dynamics of Family Relationships
	2. Internal Dynamics of Families	FCS 402 - Dynamics of Family Relationships **AND** FCS 465 - Parenting & Guidance
	3. Human Growth & Dev across the Lifespan	FCS 291- Individual & Family Development
	4. Human Sexuality	FCS 407 - Human Sexuality
	5. Interpersonal Relationships	COMS 365 Family Communication [preferred] **OR** COMS 141 Interpersonal Communication August, 2016 onward: COMS 365 - Family Communication
	6. Family Resource Management	FCS 330 - Consumer Economics

- Applicants must apply online through the CFLE-approved program application process within two years of graduation, and all courses on the checklist must have been completed no more than seven years prior to graduation.
- Applicants eligible to apply through the CFLE-approved program process DO NOT need to take the CFLE exam.

Submission Requirements for CFLE-Approved Program Application

You will apply online for CFLE certification through the CFLE-approved program process. During the online application process, here is what you will be asked to do or provide:

- Update your NCFR profile (contact, employment, and education information)
- Select your level of certification *provisional*
- Upload your completed coursework checklist from your school (filled out according to the instructions on the checklist).
- Submit an original, *official* transcript showing degree completion and successful completion of the required coursework.
- Your transcript must have a watermark, be embossed, or have other authenticating marks as described in the document. Official paper transcripts do not have to be mailed directly from the school or be in a sealed envelope. If you would like to submit an electronic transcript, it must be an official transcript and be emailed to Emily Vang from the university ("Emily Vang" and the NCFR address needs to appear as the recipient if that information is required). *Please note that photocopies of transcripts, or transcripts faxed, emailed, or printed by the applicant are NOT accepted.*
- Read and sign the CFLE Code of Professional Ethics
- Pay the application fee. Please note: Application fees are nonrefundable. (NCFR.org)

Degrees Beyond the BA/BS

To further one's education or seek specialized education beyond the undergraduate degree, students may pursue a graduate degree. The undergraduate Family and Consumer Sciences degree provides the foundation for pursuing advanced academic degrees such as a master's or PhD. Examples include:

- Individual or Family Counseling (MA)
- School Counseling (MA)
- Child Life (MA)
- Program Administration (MA)
- Family and Consumer Science Education (MS)
- College Professor (PhD)
- Master of Business Administration (MBA)

UM Accreditation, College Accreditation, Program Accreditation.

The University of Montevallo is accredited by the Southern Association of Colleges and Schools Commission on Colleges (SACSCOC). The College of Education and Human Development supports the teacher education program and the Counseling program. The Counseling program is accredited by the Council for the Accreditation of Counseling and Related Educational Programs (CACREP). The teacher education program is accredited by the Council for Accreditation of Educator Preparation (CAEP). The Family and Consumer Sciences program is accredited by the American Association of Family and Consumer Sciences (AAFCS).

AAFCS ACCREDITATION

The body of knowledge for the American Association of Family and Consumer Sciences suggests that professionals in the field of Family and Consumer Sciences relate to interrelationships among individuals, families and communities through: (a) Their study of human systems theory and life course development, (b) Capacity building, (c) Issues that foster wellness, (d) Global interdependence, (e) Appropriate use of technology, (f) Resource development and sustainability, and (g) Public policy development and analysis. An understanding of the complexity of the reciprocal relationships among individuals, families, communities, and the many environments in which they function throughout their life spans is the unifying focus of the Family and Consumer Sciences knowledge base. An understanding of individual, family, and community needs, ecosystems, and life course development underpins the dynamic, holistic, and integrative nature of the field. Following is a discussion of courses offered in the Family and Consumer Sciences program at the University of Montevallo, and how the overall focus, content, and course curriculum is rooted in the Body of Knowledge.

A core group of courses have been identified within the Family and Consumer Sciences major to further assist students in understanding the holistic and integrative nature of the field of Family and Consumer Sciences. Those courses are taken by all Family and Consumer Sciences majors regardless of concentration.

Family and Consumer Sciences Seminar focuses on the study of the body of knowledge, the integrative nature, and the conceptual foundations of the Family and Consumer Sciences profession. The course addresses the roles and behaviors of professionals in interrelationships among people. Through discussions about wellness, global interdependence, resource development and sustainability, capacity building, the use of technology, diversity, and the ecological perspective, students understand their roles as Family and Consumer Sciences professionals in the application of content to societal issues.

Introduction to Nutrition focuses on providing students with an understanding of food requirements and nutritive values of foods in order to learn how to make informed decisions about alternative food choices to meet basic human needs. Introduction to nutrition addresses the physiological requirements for food and nourishment, and students relate the interdependence of nutrition and food for maintenance and advancement of positive health, wellness, and vitality.

OMG_Studio/Shutterstock.com

Individual and Family Development.

Through observations of individuals of all ages and many ethnicities, cross-cultural assignments and classroom instruction, students are introduced to the concepts of life course development and meeting the basic human needs of individuals and families. Information about human systems theory is developed further and serves as the framework for explaining human development and behavior in the context of and in various environments. Students are introduced to the concepts of capacity building and meeting basic human needs plus safety, love, belongingness, self-esteem, and self-actualization. Course content is presented from a cultural perspective, so a student understands the impact of one's culture on development as an individual and as a family. Students enrolled in this course begin their observations at the campus Child Study Center and must follow the guidelines for developmentally appropriate practice as well as the Code of Ethics of the National Association for the Education of Young Children when they interact with the staff and children.

Consumer Economics provides an understanding to students of individual and family resource management in order to fulfill their needs and accomplish their goals which enriches community vitality in an increasingly

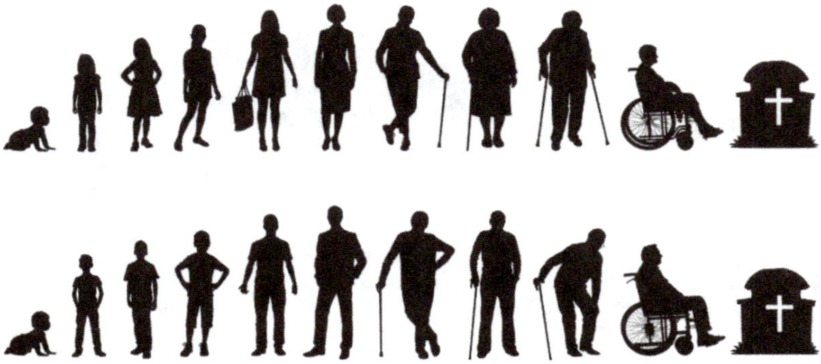

Andreas Novandi Irawan/Shutterstock.com

complex world. The Resource Development and Sustainability cross-cutting theme provides the background for students to select a topic for their resource development or sustainability project. The project assist students in understanding community vitality by researching issues that affect the environment in which the individual or family lives.

chayanuphol/Shutterstock.com

Dynamics of Family Relationships is the place students demonstrate knowledge of family strengths, including understanding, nurturing, and supporting strong, diverse, resilient families that require comprehension of complex relationships and human development across the life course. Students are able to demonstrate an understanding of capacity building of individuals, families, and communities, building on assets and strengths, respecting diversity, and responding effectively to change. Further, students demonstrate an understanding of developmental and relational aspects of family functioning. Finally, they give attention to the minority status of the family and discuss individual and family strengths that foster resilience among the characters. Dynamics of Family Relationships provides students a comprehensive understanding of the key issues facing marriages and families in the 21st century, as well as the dynamic nature of marriages, families, intimate relationships, and their diversity.

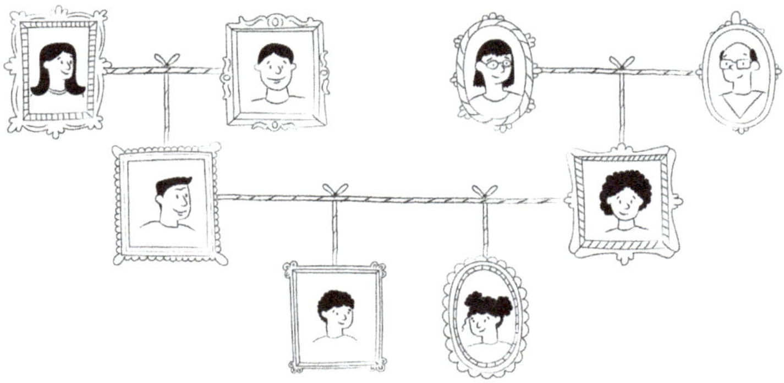

Galla Pilina/Shutterstock.com

metamorworks/Shutterstock.com

The primary focus of the **Professional and Career Development** course is the transition from school to work. The course assist students in developing skills to improve their personal and professional life by analyzing their personal strengths and weaknesses through self-assessments to promote individual well-being. Through these experiences, students begin to move up the pendulum from novice to developing professional as they learn professional protocol for social, business, and technological environments. Throughout the course, students develop an understanding of the qualities required of those in leadership positions in careers related to Family and Consumer Sciences as well as other careers. Through public meeting assignments and personal interviews, students have the opportunity to engage with publicly elected leaders. Capacity building is the theme students experience as they attend public meetings and see leaders engage with the diversity of communities, schools, and manage complex problems of today.

2

HISTORY OF FAMILY AND CONSUMER SCIENCES

HISTORY OF FAMILY AND CONSUMER SCIENCES AT THE UNIVERSITY OF MONTEVALLO

Since the beginning of the University as Alabama Girls Industrial School in 1896, education in the Family and Consumer Sciences subject matter has been an integral part of educational offerings. Programming during the early years of the University was focused on improving the living standards of society. In 1901, scientific cooking, the earliest predecessor of dietetics on campus, was offered and included instruction in basic food principles and food safety, meal management, and the purchase and safe preparation and storage of cuts of meat. Within the same year, career opportunities for young women in dressmaking and millinery were added to the Department of Domestic Science, the earliest predecessor of Family and Consumer Sciences, was formed in 1901.

By 1923, the Department of Domestic Science, renamed Home Economics in 1920 when the profession was named, had developed a four-year program and awarded Alabama College's first four-year bachelor's degree. The Department developed further its courses in quantity food production, food science, meal management, foodservice management, and dietetics. Between 1911 and 1923, an education curriculum was added to meet the growing need for adequately trained, certified teachers in public schools of Alabama. In 1923, the Department was recognized as the educational program for vocational home economics teachers by the State of Alabama.

When the profession moved to a conceptual basis in Family and Consumer Sciences with a defined body of knowledge in 1995, the program at UM moved as well. The departmental name was changed from Home Economics to Family and Consumer Sciences to reflect national trends.

Acknowledging the University's commitment to attainment of national accreditation of its undergraduate programs where recognition is available and appropriate, the Family and Consumer Sciences program has been accredited by the Council for Accreditation of the American Association of Family and Consumer Sciences since the early days of accreditation.

History of Family and Consumer Sciences at the University of Montevallo

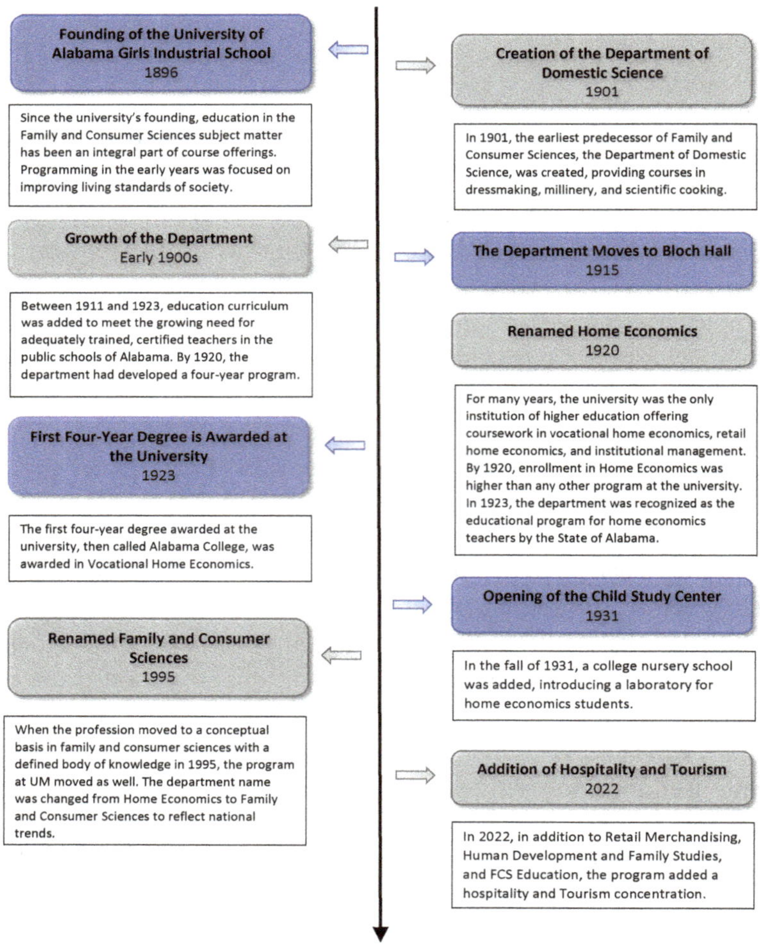

HISTORY OF THE CHILD STUDY CENTER

In 1931, the University opened a child development laboratory for preschoolers. The Child Study Center serves as a laboratory for observing children and educating future early childhood education teachers and other professionals. Opportunities for internships are also available. The play-based, full-time preschool provides developmentally appropriate experiences for children and is fully accredited by the National Association for the Education of Young Children (NAEYC). Also, it has a state funded First Class Pre-K classroom through the Office of School Readiness (OSR). Finally, it obtained a 5 STAR rating through the state of Alabama. Centers that participate in Alabama Quality STARS receive between one and five stars based on STARS Standards in the areas of Staff Qualifications and Professional Development, Management and Administrative Practices, Learning Environment and Curriculum, and Family Involvement and Community Partnerships.

The center has one operating classroom, a First-Class Pre-K classroom for 4-year old's. The classroom is set up with various learning centers including dramatic play, art, music, science, manipulatives, and writing. Classroom materials are rotated on a regular basis and meet the NAEYC standards as

Rawpixel.com/Shutterstock.com

well as OSR requirements. Children are given opportunities to learn and develop through exploration and play, and teachers have opportunities to work with individual children and small groups on specific skills. Materials and equipment spark children's interest and encourage them to experiment and learn. Planned activities are designed to help the children get better at reasoning, solving problems, getting along with others, using language, and developing other skills. There are rugs in various areas and all furniture for the children (tables, chairs, etc.) are child sized. There are safe spaces for children to go to when they feel their emotions are running too high and they need to regain their emotional or physical control. Cots are provided for each child during rest time.

A nutritious breakfast, lunch, and afternoon snack are provided daily to each child at the Center. The University and Child Study Center currently have a food service agreement with Chartwells, which is the education sector of Compass Group, the world's largest provider of high-quality food services. The Center has a Child Nutrition grant from the U.S. Department of Agriculture through their Child & Adult Care Food Program. This grant ensures that all meals and snacks are nutritious and well balanced.

Pixel-Shot/Shutterstock.com

The Child Study Center has a playground on-site. The playground is developmentally appropriate for 3–5-year old's, completely fenced in, and provides safe play structures that encourage gross motor play. In addition to the play structures, the children have access to tricycles (with helmets), various balls, hula hoops, wheel barrels, wagons, shovels, buckets, sidewalk chalk, outdoor painting, raised garden beds, a music wall, a mud kitchen and more.

3

Values and Goals

VALUES

marekuliasz/Shutterstock.com

Values, as defined by Cambridge.org, are the beliefs people have, especially about what is right and wrong and what is most important in life, that control their behavior (https://dictionary.cambridge.org/us/dictionary/english/values). Values are individual. What is most important to you, may be at the bottom of the list for someone else. Review the following terms that might represent things you value:

Faith	Education	Respect	Beauty
Privacy	Money	Leadership	Respect
Leisure time	Honesty	Reputation	Fairness
Family	Health	Safety	Serving others

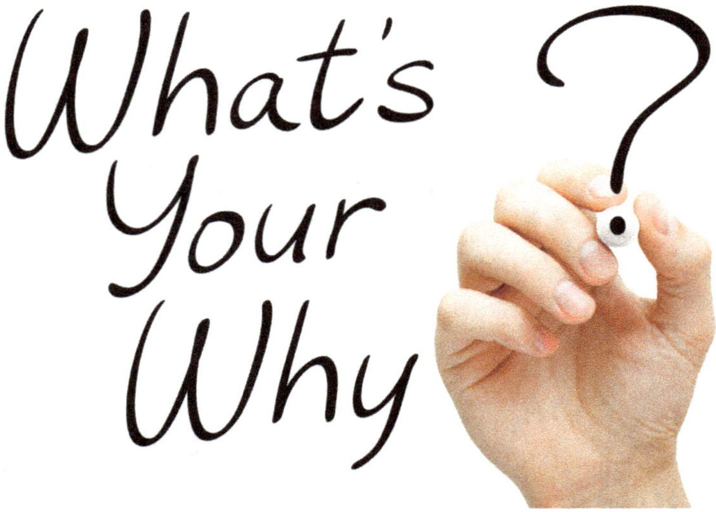

Ivelin Radkov/Shutterstock.com

So, why are values important to us? Values are what drives our decisions. Our values motivate us in our personal and professional lives. Values guide us in how we react to others and situations. Our values align with our moral code and when we do something that does not align with our values, our conscience gets to us.

ROADMAP TO GRADUATION

Road to success:

wenich_mit/Shutterstock.com

Your success is dictated to a great extent by your daily routine. These tips may sound repetitive, but with more than 30 years of combined teaching experience at the collegiate level, the

Ihor Biliavskyi/Shutterstock.com

faculty can attest that students have more success when they incorporate these tips into their college plan. Not only will these skills assist you in being successful in college, but in life as well. As you begin your journey and plan your goals, remember the acronym SMART – specific, measurable, attainable, relevant, time-limited.

- Goal Setting
  - A key component of goal setting is using positive language. For example, use phrases such as "having a GPA of 3.0 or above" instead of saying "trying not to fail a class this semester".

BonNontawat/Shutterstock.com

  - Specific – State exactly what you want to achieve today, this week, this month. For example, instead of saying, "I am going to study this week." Restate it by saying, "I am going to study 3 hours per day."
  - Measurable – How will you know if you have been successful if you cannot measure it. Therefore, track your time daily of how long you study. Set your watch and measure the time.

  - Attainable – Include goals that are realistic and that you have sufficient time to achieve. If you work and go to school, be realistic with your time. If you have not previous planned time for studying, then start with a small goal that is achievable and increase that goal each week.
  - Relevant – As you set goals, make sure they are relevant to your success. As you are a student, going to class and studying is relevant. If you must work part-time, you will need to set a limit on the number of hours you will work in order to be successful at attending class and studying.
  - Time-limit – You should be able to achieve your goals in a timely manner. Perhaps your long-term goal is to graduate with your degree in four years. You can break this long-term goal into specific, measurable, attainable, relevant goals by semester. This will give you a sense of accomplishment and success in smaller chunks.
- Understanding individual learning preferences
  - Everyone learns differently. It is important to understand how you learn best so that you can engage in course lectures, course material, assignments, and examinations in the most effective manner for you individually. How do you learn the best? Are you a
    - Tactile Learner – hands on learning. This may be someone that highlights texts, takes notes, and creates projects.
    - Kinesthetic Learning – life experience learning; role playing, interactive scenarios, movement.
    - Auditory Learner – prefers to listen to lectures, engage in discussions, may repeat information aloud
    - Visual Learner – prefers to read material, see and study pictures, may color-code notes, create diagrams or mind maps, and watch videos.
- Be organized
  - As a student, you may have already determined that college is very different from high school. Your course schedule may include courses only on Monday, Wednesday, and Friday, leaving Tuesday and Thursday as "free" days. College professors will typically lecture during class and expect you to take your own notes. The question is what will you do with these notes? It is critical for long-

term success that you keep your personal and academic materials organized.

- Balance (school, social, and work)
  - Manage your schoolwork load (schedule out time for each project, and time to study for quizzes and exams). Use a calendar to schedule this time.
  - Do not wait until the last minute – you never know what will happen the night before a project is due.
  - Time management skills and scheduling will assist in planning your time for school, work, and fun. Treat school as you would a job. Get up early, eat a balanced breakfast and start your day at 8:00 a.m. Locate a place you feel comfortable while studying, preferably one with minimum distractions. Read course material, work on an assignment, or study for an exam until your first class of the day. Attend class, then return to your course work until the next course. If you continue this pattern until approximately 5:00 p.m., you should now be ready for the fun part of your day.
- Show up for class (attend and be present in class)
  - Be aware of professors' attendance policies. Attendance policies may be found in the syllabus for each individual course and may vary by professor and course. Attendance in class or non-attendance in class may impact your final grade.
    - Utilize campus resources

ORION PRODUCTION/Shutterstock.com

    - Learning Enrichment Center
    - Harbert Writing Center
    - Access and Compliance Office
    - Financial Aid Office
    - Counseling Center
    - Career Development Center
    - Carmichael Library
- Make time to speak with your professors and advisor
- Build healthy study habits
  - Treat school as if it's a job.
  - Form a study group
  - Turn your devices off
  - Most students study by rereading their textbooks or notes. This is not the most effective way to study. It is time consuming and does not result in mastery of the content. Instead, students should practice retrieval, space out practice, and mix up studying (Brown et al., 2014).
    - Practice retrieval: Rather than rereading or highlighting notes, attempt to recall information from your memory. Ho through the chapter and the lectures and write open-ended questions related to the topics. You can also use questions from previous quizzes and from the textbook. When quizzing yourself, pay attention to what you don't know. Additionally, studying with other people has been proven to be a highly effective means of studying when you and your peer (1) ask each other questions, (2) explain concepts as if you are teaching them to the peer, and (3) make connections between course concepts.
    - Space out practice: Leave considerable time between practice sessions (multiple days if possible). To do this, you will need to start practicing early and often. It is beneficial to set some time aside each week to write questions and practice retrieval. As you are practicing new material, connect back to previous material.
    - Mix up your studying: You can mix up your studying in a few ways. First, treat your practice sessions like a test and don't go chapter by chapter or topic by topic. Second, while you are spacing out your practice, study content from other courses.

- Other effective methods:
  - Elaboration: Relating material to what you already know, explaining it to somebody else in your own words, explaining how it relates to your life
  - Generation: The attempt to answer a question or solve a problem before being shown the answer. The act of filling in a missing word results in better memory of the text than simply reading a complete text. Experiential learning is an example of this. As you are reading new material, try to explain beforehand the key ideas you expect to find and how you expect they will relate to previous material. For math or science classes, one example would be trying to solve new problems before you get to class.
  - Reflection: The act of taking a few minutes to review what has been learned in a recent class or experience and asking yourself questions. What knowledge or other experiences does this remind you of? What might you need to learn or need more clarification on?
  - Calibration: Using an objective measure to clear away any illusions of knowledge. The aim is to be sure that your sense of what you know and can do is accurate. Quizzes are a good way to do this
  - Mnemonic devices: "Please excuse my dear Aunt Sally." Tools to create mental structures that make it easier to retrieve information.

- Note-taking skills: use active listening during class, write down phrases versus full sentences (abbreviations and/or shorthand are acceptable), engage in discussions to reinforce content, review notes shortly after class, highlight important concepts, and make note of any questions you may want to go back to.
- Form relationships with peers

College life allows ample time to develop healthy relationships and friendships with others. Human nature finds us hanging out with those most like us. Try to step out of your comfort zone and enjoy the diverse mix of people on campus. Realize that not all relationships and friendships will

Brown, P. C., Roediger, H. L., & McDaniel, M. A. (2014). Make it stick: The science of successful learning. Belknap Press.

last forever. It is perfectly alright to make many acquaintances along the way. If you are more of an introvert, hang out with those that might be more adventurous than you would be normally. Saying that, it is equally important that you are able to recognize when a friendship or relationship is unhealthy. When friends and acquaintances are not encouraging you to reach your goals or they are trying to change your morals and values, remove those toxic people from your circle.

- Get involved on campus
  - Greek life
  - Campus organizations
  - Departmental student organizations
  - Study Groups
  - Homecoming activities
  - Volunteer on campus during events
  - Work on campus

Rawpixel.com/Shutterstock.com

- Take care of yourself

  Although your college experience will end, you will always be you. Taking care of yourself is important to your overall well-being. As you will learn in the Family and Consumer Sciences courses, individual and family well-being and community vitality are goals of the program. When individuals have a positive outlook, they perform better as members of a family and perform better in the workplace.

  - Make choices you will be proud of later
  - Stay focused on your goal
  - Eat well
  - Exercise
  - Get enough rest
  - Take time out for yourself

CleanVector/Shutterstock.com

TIME TO GET STARTED

- Plan a meeting with your advisor to discuss your major and plan of study. Students should meet with their advisor a minimum of once per semester.
- Understand your plan of study and the course progression.
- Keep records of your meetings.
- Review your degree plan on DegreeWorks.

Media_Photos/Shutterstock.com

PROFESSIONALISM

Professionalism skills are learned behaviors and competencies that demonstrate respect, responsibility, and positive interaction in the workplace. They encompass a wide range of abilities, including communication, teamwork, problem-solving, and strong work ethics. Mastering these skills is crucial for career advancement and success. Professionalism includes your character, your knowledge, your experiences, and your image. Let's talk about your character. Your character is who you are. Who you are is separate than what you are. What you are speaks to your standing in a family (wife, husband, brother, sister, mother, etc.), your work (preschool teacher, doctor, fireman, mechanic, dental assistant, customer service associate, etc.), and your position (CEO, manager, associate professor, etc.). Who you are is a much deeper level than just what you are. Who you are gets to the very foundation of what makes you: values, morals, beliefs, spirituality, intellect, culture, physicality, social, emotions, dreams, etc. Take a minute to think about who you are. What are your strengths and growth areas in each of the areas listed above. What terms would you use to describe your strengths: for example, versatile, trustworthy, analytical, decisive, accountable, ethical, optimistic, self-reliant, warm, grounded, resourceful, etc. These are just a few terms that might tell someone who you are. Now that we are on the way to understanding who we are, the next question is what do you want from life and work?

There is an old saying, "If you don't know where you are going, how will you know when you arrive?" Have you asked yourself, what will it take to make me truly happy, successful, and provide a since of fulfillment. Consider the list of questions below:

- Am I truly motivated by my current career choice?
- What majors will lead me to be successful in my career?
- Is my career more important than my family and friends?
- Are my family and friends more important than my career?
- Do I like working inside or outside?
- Do I want to travel for work?
- Am I looking for a career or a job?
- Is salary more important than the work environment?
- What am I willing to do to get ahead?
- What is my passion?

PROFESSIONAL COMMUNICATION SKILLS

Communication includes verbal, non-verbal, written, and visual. Verbal communication includes the use of language both written and spoken to articulate thoughts and ideas clearly and concisely to others. During verbal communication, one should be mindful of the tone they are using. Also, during spoken communication, active listening should be employed. For effective communication to occur, the message must be delivered, received, and understood.

The need for effective communication skills go beyond one's relationships with their families and friends. Effective communication skills are important to one's success in school and the workplace. A study in 2022 by Grammarly and The Harris Poll found that U.S. businesses annually lose an estimated $1.2 trillion due to ineffective communication. (Businesswire.com February, 22, 2023). Also, the study revealed that with poor communication, businesses see lack of productivity, missed deadlines, and rising cost just to name a few. In school, it is important to communicate with your peers, staff, professors, and administrators concisely.

In written communication, include all relevant details, make sure details are correct, use the appropriate level of formality, check spelling and grammar, and ensure that the message is understandable.

E-MAIL: COMMUNICATION

one photo/Shutterstock.com

Sending an email to one of your professors might be intimidating, especially if you do not know them very well. To help overcome this issue, below are best practices to consider.

- Emails are more formal than texting, Snapchat, and other digital media.
  - Emails should have a clear subject line:
  - FCS 100 assignment question
  - Request for advising appointment
- Always begin with a polite salutation, Dear Dr. Smith or Hello Dr. Smith. It will be important to check the syllabus to find the correct title for your professor. Addressing faculty, staff, and administrators at a university as Bro or Dude is not appropriate.
- Always introduce yourself even if the professor knows who you are. Remember, professors teach several classes each semester, some with large numbers of students in each class. Example, My name is Jane Smith, and I am in your FCS 100 class.
- State your situation in a clear and concise manner. This is not a text, and LOL or BTW are not acceptable vocabulary words in an email.
- Check for spelling and grammatical errors. You want to be taken seriously, so it is a good idea to write in a professional manner.
- Use a formal closing: Sincerely, Thank you, Best regards

Before sending your email, do your part to try and solve the question by checking the syllabus or Canvas for answers if it is a question related to class or a class assignment. Inappropriate email content such as "I really need to pass this class with a A," or "I forgot about the homework, will you take it late?" are not the professor's issues, they are personal issues which you should take responsibility for.

RESUME WRITING

Prathankarnpap/Shutterstock.com

Resumes should be considered a piece of professional communication. The resume represents you through written communication. A resume is an extension of your brand image. What you share and how you share your educational background, work experiences, extra-curricular activities, and other qualifications speaks volumes to potential employers. A resume is the first piece of communication you send to potential employers and will determine if you move on to the next phase of the employment process. Each time you submit a resume for a potential job, you should tailor it to that specific job posting. Remember, resumes are not a one size fits all.

There are numerous formats and styles for resumes. Each individual must determine the style and format that best suits one's experiences. Individuals just finishing their college degree and beginning their career should typically limit their resume to one page. After you have been working in your career for many years, you may increase the length of your resume.

When beginning to develop your resume, you should consider the job description. What are the qualifications and characteristics that the employer mentioned in the job posting? Notice keywords listed in the job description and include those in your resume. The resume should not be written in narrative form nor include the term "I".

Each resume should include the following:

- Contact information – name, address, telephone number, and email address
  - Your email address should be an appropriate email address (dw.bell@school.edu)
- Education
- Work experiences
  - These should be relevant work experiences to the position you are applying
  - These may be paid or unpaid experiences (example: internship)
  - Bullet points that showcase how or why you did something
    - Developed lesson plans by using developmentally appropriate activities for 4-year-olds.
    - Increased sales for the department by analyzing past sales trends and purchasing merchandise to meet target market wants and needs.
- Specific qualifications and certifications
  - Expertise in Excel spreadsheets
  - CFLE certification
  - FCS State Teaching Certification (grades 6-12)

INTERVIEW SKILLS

Once you have received notification from a company that they would like to schedule an interview, you should begin preparing. Do not go into an interview and think that your educational degree alone will secure the job. There will be many people with the same education degree applying for the same position. How will you stand out among them? Remember, that everyone you meet at the company is a chance for you to make a good impression.

BeeBright/Shutterstock.com

- **Prepare/Roleplay**
  - Research the culture, values, and mission statements of the company for which you will be interviewing. If the company has an online presence, review what is shared with the public.
  - As you prepare to answer questions and ask questions, remember this tip – the company is not interested in what it can do for you, rather the interviewer is interested in what you can do for the company.
  - Prepare a list of questions that you would like to ask during the interview. It is advisable to make these questions about the company. This is not the time to ask about salary or benefits.
  - Although you have submitted a resume previously, you should bring several copies of your resume with you. The resumes should be in a professional portfolio (leather binder). You should not carry loose papers into an interview, nor should you carry a backpack into an interview.
  - Role play with a family member or friend. Have a list of general questions that the individual can ask you that are asked in most interviews. Have the individual write down your responses or record the interview. This will provide you with areas to work on prior to

the actual interview. Pay close attention to your grammar during the practice session.
  - Tell us about you.
  - What are a few of your strengths?
  - What are a few of your growth areas?
  - Tell us about a time you had to work as part of a team, whether in school or at a place of work?
  - How do you handle conflict?

Pixel-Shot/Shutterstock.com

- Prepare your interview outfit well in advance of the actual interview.
  - Research the company and find out who you will be interviewing with if possible.
  - Keep in mind that most hiring managers are not recent college graduates; therefore, may have a more conservative expectation of appropriate interview attire. Remember, each company may have a dress code for employees once you are hired, but you are interviewing for a position, not an employee. It is up to you to make a good first impression and your appearance will say a lot about you.

Shutterstock AI

        - Make sure the outfit is clean, free of stains and smells, and pressed well. If you are a smoker, it is best if the outfit does not smell like smoke or heavy fragrance.
    - Make sure you know the exact location of the interview and how to get there. If you are unfamiliar with the location, making a practice run the day before at the same time of day you will need to travel is an excellent idea. It is advisable to have a full tank of gas and know where you will park.
    - Get plenty of sleep the night before.
- **The Day of the Interview**
    - Get up early and eat before you go to the interview.
    - Getting ready
        - Ensure your outfit is clean, pressed, and free of smells (smoke, heavy fragrance, etc.)
        - If you are a smoker, do not smoke in the car on the way to the interview. The smell of cigarette smoke attaches to your clothes and can be a turn-off for many people.
        - Shoes should be cleaned and appropriate for an interview. Closed toe shoes are recommended for both male and females along with appropriate leg coverings (socks and hosiery). You may not be accustomed to wearing socks and hosiery daily, however, this is an interview, and it is expected.
        - Good grooming habits are a must along with fresh breath.
        - If you have long hair (male and female), pull it back away from your face. This prevents nervous hands from playing with hair.
        - Consider removing facial piercings before the interview.
        - Tatoos are more acceptable today; however, for an interview it is better to air on the side of caution and cover visible tattoos. You may not be aware of a company standard for visible tattoos. If offered a position, this is an area that you would need to discuss at that time.
        - Bring your leather portfolio with your prepared resume copies.
        - Bring a notepad to jot down questions during the interview.
        - Do not chew gum during an interview.
        - Do not bring food or drink to an interview.
        - Do not bring anyone into the building with you for an interview.

  - o Turn off all electronic devices, including notifications on your watch prior to entering the building for an interview. Turn the devices completely off to prevent any distractions during the interview.
  - o Arrive at least 15 minutes early for the interview. Plan for delays and unexpected events.

- **During the Interview**
  - Enter the interview with confidence and a positive attitude.
  - Greet the interviewer with a firm handshake and good eye contact.
  - Be conscience of your non-verbal communication such as posture, eye contact, nervous habits, fidgety hands and legs, etc.
  - Answer each question to the best of your ability and thoroughly. Providing brief answers does not allow the interviewer to get to know you and your ability. It is ideal to have a conversation rather than a question/answer session.
  - Take notes of things that you would like to follow-up with later in the interview.
  - Be conscience of your grammar. Speak clearly and enunciate your words.
  - It is never a good idea to degrade your previous place of employment or school you attended. As the saying goes, if you can't say anything good about someone, don't say anything at all.
  - When it is appropriate, normally at the end of the interview, the interviewer will ask if you have any questions. This is the time to ask any follow-up questions you jotted down during the interview as well as any you prepared prior to the interview.
  - Avoid questions that concern salary, benefits, work hours, vacation, flexible schedules, and so forth. These questions are only important if you are offered a position. At that time, you will have the opportunity to discuss and negotiate.
  - Thank the interviewer(s) for their time and the opportunity to interview for the position. Shake hands with each person in the room and collect a business card if possible.
- **Follow-up**
  - After the interview, follow-up with a personalized handwritten thank you note to each person that you interviewed with.

4

Careers in Family and Consumer Sciences

CAREER DEVELOPMENT

Career development does not begin or end when you get your first job. Your time spent in the FCS program will help you to start your career development journey. The faculty will assist you in aligning your values, passions, and strengths with potential career choices, and the internship in your senior year will provide you with valuable hands-on experience. One of the most important tasks as a student is understanding what it takes to reach your career goals and working towards completing those requirements. For some, this may include additional certifications or degrees following graduation. For others, your career might require you to attend continuing education events (e.g. conferences, workshops, and webinars) to ensure that you are up to date on current trends and research. As you are considering career opportunities, be diligent in researching these requirements and identifying gaps in your skills and knowledge.

CAREERS IN FAMILY AND CONSUMER SCIENCES (FCS)EDUCATION

- FCS Educator (Secondary Grades 6-12)
- FCS University Faculty
- Curriculum Developer
- Corporate Trainer

- Human Resources
- Cooperative Extension Agents
- Family Support Advocate
- Entrepreneur
- Health and Wellness

CAREERS IN HOSPITALITY AND TOURISM

- Event Planner/Manager
- Sports Center Manager
- Hotel Management
- Restaurant Management
- Sales/Marketing
- Food and Beverage Manager
- Food Purchaser/Broker
- Spa Manager
- Tour Guide/Operator
- Travel Agent
- Culinary Arts Professional
- Events Coordinator
- Cruise Director
- Recreation Director

CAREERS IN HUMAN DEVELOPMENT AND FAMILY STUDIES

- Early Childhood Educator
- Early Childhood Administrator
- Family Policy Specialist
- Certified Family Life Educator
- Youth Program Director
- Senior Activities Coordinator
- Early Interventionist

- Behavioral Specialist
- Human Service Worker
- Case Manager
- Family Resource Coordinator

CAREERS IN RETAIL/FASHION MERCHANDISING

PGorodenkoff/Shutterstock.com

- Product Development
- Fashion Designer
- Fashion Director
- Stylist
- Social Media Specialist
- Visual Merchandiser
- Merchandise/Fashion Coordinator
- Retail Buyer
- Trend Forecaster
- Store Owner
- Store Management
- Fashion advertising
- Fashion Journalism
- Marketing/Sales

ncfr
catalyzing research, theory and practice
CAREER OPPORTUNITIES in FAMILY SCIENCE
Business, Consumer & Family Resource Services
Employee Assistance
Corporate Childcare
Family Financial Counseling & Planning
Vocational Guidance
Consumer Protection
Family Resource Management
Food Assistance
Child and Family Poverty Research
Work Life Balance
Family Business Consultation
Academic Advising
Community-Based Social Services
Youth Development Programs
Adoption Agencies
Foster Care
Teen Pregnancy Prevention & Assistance
Family Preservation
Welfare Assistance
Vocational Rehabilitation & Job Training
4-H Youth Development
YMCA/YWCA
Scouting Programs
After-school Programs
Adult Day Care
Gerontology Programs
Education
Public Schools
Cooperative Extension
University & Colleges
Research
Family Life Education
Community Education
Parent Education
Family Peace & Justice Education
Children's Museum Education
Marriage & Family Enrichment & Education
High School Guidance Counseling
Program & Curriculum Development
College Student Life Programs
Family Intervention
Individual & Family Therapy
Case Management
Crisis & Hotline Services
Court-Mandated Parent Education
Divorce Mediation
Abuse Protection Services
Sexual Violence Prevention
Drug & Alcohol Prevention
Residential Treatment Programs
Victim/Witness Support Services
Faith-Based Organizations
Clergy
Family Mentor
Parent Education
Youth Programs
Budgeting and Money Management
Marriage and Relationship Programs
Early Childhood Education
Childcare Centers
Head Start Programs
Preschools
Early Childhood Family Education
Government & Public Policy
Family Policy Analysis
Advocacy/Lobbying
Military Family Support Services
Child & Family Services
Juvenile Justice
Family Court
Healthcare & Family Wellness
Public Health Programs & Services
Child Life
Nutrition, Wellness and Sexuality Education
Prenatal and Maternity Services
Disability Services
Long-Term Care Programs
Hospice
Funeral Services
International Education & Development
International Family Policy Analysis
Peace Corps and NGO Leadership
Global Family Planning Programs
Community & Sustainable Development
International Human Rights Advocacy
Immigration & Migrant Families Services
Research
Grant Proposal Writing
Academic and Government-Related Research
Population Studies & Demographic Research
Research for Community Non-Profit Agencies
Program Evaluation & Assessment
Writing & Communication
Curriculum & Resource Development
Public Service Radio and TV Programming
Newspaper & Magazine Journalism
Blogging & Social Media
National Council on Family Relations • ncfr.org

www.ingramcontent.com/pod-product-compliance
Lightning Source LLC
LaVergne TN
LVHW022321200926
848481LV00004B/1397
* 9 7 9 8 3 8 5 1 9 4 1 0 0 *